THE SUNFLOWER EXPERIMENT

Written By
Teresa Peck

Illustrated By
Whimsical Designs By CJ

Printed in the United States of America

ISBN: 978-0-578-76364-4

To my parents, to whom I owe everything.
To my daughters, Shali & Jameela, for always encouraging me to write.
And to the students Of Hardy Williams Mastery Campus in Philadelphia

Aa Bb Cc Dd Ee Ff G

Lily was excited about today's science class. Ms. Thomas was teaching about seeds and the baby plants inside of them.

"Does every seed have an embryo Ms. Thomas?" asked Lily.

"Yes Lily," replied Ms. Thomas. "Every seed has a baby plant inside of them called an embryo." Lily smiled at the thought of a baby plant.

Ms. Thomas reminded the class about the rules of the science lab. She pointed to the chart on the wall. "What is rule #1?" she asked.

“Do not eat or drink anything in the lab,”

all the children replied together.

Lily recalled the rainbow candy experiment from last week. She and her partner Michael arranged brightly colored candy in a rainbow pattern.

The candy looked delicious. It came in different flavors: strawberry, grape, lemon, and apple. Lily wanted to sneak a piece of candy. No one will just miss one tiny piece of candy she thought.

As Lily reached to grab a piece of candy Michael grabbed her hand to stop her. "Don't Lily!" Michael exclaimed. "You know the rules. Besides you'll mess up the experiment."

Ms. Thomas began passing out seeds to each set of partners. Tamica and Tara got bush bean seeds. Niecy and Jermain got pea seeds. Nicky and Charlie got popcorn seeds, and Lily and Michael were given sunflower seeds.

Lily shouted, "These seeds don't have baby plants inside of them. They're sunflower seeds. I buy these at the store and eat them all the time."

"Yes!" Ms. Tabourn said.
"But remember! We don't eat anything we use for an experiment."
All of the children continued writing in their science journals:

'What effect does water have on seeds?'

Lily remembered the last experiment. She ate a piece of candy and nothing happened so she decided to eat a sunflower seed. Lily opened the shell and popped a seed in her mouth. Science class ended moments later.

The students cleaned up their stations and lined up for lunch.

When Lily arrived home later in the afternoon, her mom had a snack waiting for her.

"Hi Lily."

"Hi Mom."

"How was your day?"

"Boring," Lily moaned.

"Did you learn anything new?" her mother asked.

"Ms. Thomas is teaching us that all seeds have embryos. An embryo is a baby plant."

"Is that so?" her mom asked.

"Yes, but when I opened a sunflower seed, I didn't see an embryo. It was just a seed. I eat those all of the time."

"Did you eat any seeds in class?"

Lily thought her mom would be upset if she knew she ate a seed in class. "No mom," she replied.

Lily finished her snack and drank a cup of water.

She grabbed her backpack then left to start her homework.

Lily time for bed," her mother called to her.
"Take a bath and I'll be up to tuck you in."

Lily was settled in her bed. Her mother tucked her in and turned out the light. "Good night Lily."

"Good night mom," Lily whispered.

Lily had trouble falling asleep.
She knew she should not have eaten the seed.
"My seed didn't have a baby plant inside of it she thought. It was ok to eat. Lily nestled in her bed and drifted off to sleep.

Lily had trouble staying asleep.

It was just a bad dream.

The next morning, Lily rushed to class to talk to Ms. Thomas.
"Ms. Thomas! I have something to tell you," Lily exclaimed.

"What is it?" Ms. Thomas asked.

Lily stuttered, "I . . . I ate a sunflower seed in yesterday's class. I thought it would be ok because it didn't have an embryo inside." But then I had a bad dream and I turned into a sunflower!"

"Oh Lily you're shaking!" said Ms. Thomas right before she hugged Lily. "It's ok. I'm glad you told me and don't worry. You won't turn into a sunflower. The seeds won't grow because you ate them. Birds eat seeds all the time and they don't turn into flowers."
"Why not Ms. Thomas?"

"All seeds have an embryo inside of them. When the seed soaks up water, it begins to change.
The seed swells and begin to grow a new plant."

Lily begins to smile at the thought of a baby plant growing into a new plant.

Lily hugged Ms. Thomas and promised never to eat anything again in the science lab.

"I'm glad Lily. Now hurry up, class is about to begin."

Lily unpacked her backpack and sat next to Michael.

"Are you ok?" asked Michael.

"I am now," replied Lily.

"Today we will go on a seed hunt class," said Ms. Thomas.
She asked the students, "Do different kinds of fruits, have different amount of seeds?"

Lily's eyes got big.

Ms. Thomas showed the class a tray of fruit: strawberries, apples, oranges, cherries.

"Here we go again," Michael sighed.

Lily smiled.

Ms. Thomas smiled back.

Procedure to Conducting The Candy Investigation

Skittles Science Experiment

1) Make a circle with the Skittles in a single row colored pattern around the edge of the plate.

2) Slowly pour enough warm water to cover all the Skittles and the plate itself.

3) Wait and watch what happens.

Made in the USA
Middletown, DE
19 December 2020

29387736R00015